The Father of Modern Rocketry:

The Life and Legacy of Robert Goddard

Writer: A. Scholtens

Cover design: A. Scholtens

© A. Scholtens

January 2023

Preface

Robert Goddard

American engineer, professor, physicist and inventor

In this book, we delve into the life and legacy of Robert Goddard, widely considered the "Father of Modern Rocketry." Goddard's pioneering work in the early 20th century laid the foundation for the development of rocket technology and paved the way for space exploration.

Goddard's passion for science and technology began at a young age and continued to drive him throughout his life. Despite facing skepticism and lack of funding from the scientific community, Goddard persisted in his research and experimentation.

Through this book, readers will learn about Goddard's early life and education, his groundbreaking work in rocket technology, and the impact of his contributions on the field of space exploration. We also examine Goddard's personal life and the challenges he faced during his career.

This book is an in-depth look into the life and legacy of Robert Goddard, a true pioneer in the field of rocketry and space exploration. His contributions to science and technology have had a lasting impact on the world and continue to inspire future generations of scientists and engineers.

A. Scholtens

January 2023

Introduction

Robert Goddard, known as the "Father of Modern Rocketry," was an American engineer, professor, physicist, and inventor who is credited with creating and building the world's first liquid-fueled rocket. Born in 1882 in Worcester, Massachusetts, Goddard's early interests in science and technology would ultimately lead him to make groundbreaking contributions to the field of rocket science. He spent much of his life researching, experimenting, and developing new technologies that would pave the way for the future of space exploration.

Goddard's work was not without its challenges and obstacles. Despite facing skepticism and criticism from many in the scientific community, he persisted in his research and experimentation, driven by his passion for understanding the principles of rocket propulsion. His relentless pursuit of knowledge and innovation would ultimately lead to the launch of the world's first liquid-fueled rocket in 1926, a historic event that would forever change the course of rocket science and space exploration.

Goddard's legacy lives on today, as his pioneering work in rocketry continues to inspire scientists and engineers around the world. His contributions to the field of rocket science, including his development of the liquid-fueled rocket, his work on rocket stabilization and guidance systems, and his research on the potential for rocket propulsion in space, have played a crucial role in the development of modern space technology.

This book will explore the life and accomplishments of Robert Goddard, delving into the details of his early life and education, his groundbreaking work in the development of the liquid-fueled rocket, and the launch of the world's first liquid-fueled rocket in 1926.

It will also examine Goddard's later years and his lasting legacy in the field of rocket science and space exploration. Through this comprehensive examination of Goddard's life and work, readers will gain a greater understanding of the man and the revolutionary impact of his contributions to the field of rocketry.

Table of Content

Chapter 1: Early Life and Education

Robert Goddard was born on October 5, 1882, in Worcester, Massachusetts, the eldest son of Nahum Danford Goddard and Fannie Louise Hoyt Goddard. His father was a tobacco and candy merchant, while his mother was a housewife. Nahum was a hardworking man and a strong supporter of his children's education, he also had a great interest in science and technology and this had a great impact on Robert's interest in science and technology. Fannie was a loving and supportive mother, who encouraged her son's curiosity and provided him with the books and materials he needed to pursue his interest in science and technology.

Robert had five siblings: Esther, Richard, Ethel, Abigail and Sarah. Esther, the eldest of the sisters, was a hardworking and responsible person, who helped her mother in raising the children. Richard, the second eldest brother, was a hardworking man and was also interested in science and technology. Ethel was the third eldest sister, she was a kind and caring person and supported Robert in his early experiments and inventions. Abigail was the fourth eldest sister, she was a talented artist and helped Robert with his drawings and illustrations for his research papers. Sarah, the youngest of the siblings, was a bright and curious child, who was inspired by Robert's work and later became a scientist herself.

Robert Goddard's family was a close-knit and supportive family, who encouraged his interests and provided him with the support and resources he needed to pursue his passion for science and technology.

During his childhood, Robert Goddard was an avid reader and had a strong interest in science and technology. Some of the books he read in his youth were Jules Verne's "From the Earth to the Moon" and "Around the Moon," which sparked his interest in rocketry and space travel. He also read books by H.G. Wells, such as "The War of the Worlds," which further fueled his imagination and curiosity about the possibilities of science and technology. Additionally, Goddard read books on physics and mathematics, which helped him to develop a deeper understanding of the principles of rocket propulsion.

In his youth, Goddard also conducted experiments in his family's backyard. Some of his early experiments involved launching small rockets made of bamboo and paper, and measuring their performance. He also experimented with different types of fuels, such as gunpowder and gasoline, to see which would be the most efficient and stable source of propulsion. Goddard also built and launched small balloons to study the effects of altitude on various materials and to understand the concept of lift. These early experiments were crucial in developing his understanding of the principles of rocket propulsion, and they laid the foundation for his later work in rocket science.

In his early years at South High School in Worcester, Goddard competed in several math and science competitions and won several prizes. For example, he participated in the mathematics competition of the Mathematical Association of America and won first prize. He also entered the American Physical Society competition and won second prize. His math and science teacher at South High School, Mr. George W. Walker, described Goddard as one of the most brilliant students he ever had, and that Goddard had "a natural aptitude for math and physics, and a keen interest in them". His outstanding achievements in high school helped pave the way for his later success in rocket science and space exploration. He graduated at the top of his class in 1900.

Robert Goddard chose Worcester Polytechnic Institute (WPI) for his undergraduate studies because of its reputation for providing a strong education in the fields of engineering and science. WPI, founded in 1865, was one of the first polytechnic institutes in the United States and was known for providing students with hands-on, practical training in engineering and technology. This was a perfect match for Goddard's interests and passions, as he was already interested in science and technology and had been conducting experiments in rocketry since his youth. Additionally, WPI was located in Worcester, his hometown, making it a convenient choice for him.

After completing his undergraduate studies at WPI, Goddard continued his education at Clark University in Worcester. Clark University had a strong graduate program in physics and a reputation for being an institution that fostered independent research. At Clark, Goddard was able to study under the guidance of several prominent scientists, including physicist Ernst Mach, who proposed that the speed of sound was not a fundamental constant, but rather depended on the density of the air. This idea would later inspire Goddard's research on the use of rocket propulsion in space, where the air is much thinner.

Additionally, Clark University was known for its research in the field of atmospheric physics, which was closely related to Goddard's interest in rocket propulsion. The University's facilities and resources, combined with the opportunity to work with respected scientists and researchers in his field, made Clark University an ideal choice for Goddard to pursue his graduate studies.

After completing his graduate studies, Goddard began to focus on the study of rocket propulsion. He conducted a series of experiments in which he fired small rockets and measured their performance, and he began to develop new technologies that would improve the efficiency and stability of rockets.

He faced skepticism and criticism from many in the scientific community. In 1919 he published a paper titled "A Method of Reaching Extreme Altitudes" in which he described his ideas for using rockets to reach the upper atmosphere and even outer space. In the paper, Goddard suggested using liquid fuels, such as liquid oxygen and gasoline, as a more efficient and stable source of propulsion. He also discussed the potential of rocket propulsion in space travel and suggested a number of ideas for future research, including the use of rockets for scientific research and the exploration of the upper atmosphere.

However, many in the scientific community were skeptical of Goddard's ideas and criticized his paper for what they saw as unrealistic and fanciful ideas. For example, on January 13, 1920, The New York Times published an editorial written by its science editor, Dr. A. Lawrence Lowell. He was highly critical of Robert Goddard's work in rocket technology. The editorial, titled "Aerial Navigation: A Cold Fact and a Colder Doubt", argued that Goddard's work was "purely visionary" and that it was unlikely that a rocket could ever reach the upper atmosphere, let alone outer space. The editorial also claimed that Goddard's work was misleading and that he had not understood the basic principles of physics.

Dr. Lowell wrote the editorial after reading Goddard's publication "A Method of Reaching Extreme Altitudes". Lowell was skeptical of Goddard's claims, arguing that a rocket could not travel through the vacuum of space because there would be nothing to push against. He also stated that Goddard's ideas were based on an incorrect understanding of the physics of rocket propulsion.

The editorial generated much controversy and was criticized by many scientists and engineers for its inaccuracies and lack of understanding of Goddard's work. Goddard himself was deeply hurt by the editorial, feeling it had damaged his reputation and hindered the progress of his investigation.

It's not entirely clear why Dr. A. Lawrence Lowell, the science editor of The New York Times, believed he had the authority to disqualify Robert Goddard's work. However, it is likely that as a science editor he felt he had a certain level of expertise and authority when it came to evaluating scientific ideas and theories. In addition, Lowell may have felt that Goddard's work was inconsistent with the scientific understanding of the time and unlikely to be successful, so he felt compelled to share his views with the public.

It is also possible that Lowell, who was a Harvard professor, had some degree of bias, given that Goddard's research was not conducted in a well-known institution and it lacked the support of the scientific community, Lowell would have dismissed Goddard's work if not credible.

We should also remember that Goddard's work was not well known at the time and was met with much skepticism from the scientific community. Lowell's editorial was not an isolated incident, but rather a reflection of the general attitude towards Goddard's work at the time.

However, it is important to note that Lowell's opinion was not based on a thorough examination of Goddard's work and that he was not an expert on rocketry and space exploration.

As we now know, Goddard's work was actually ahead of its time, and many of his predictions and ideas proved correct in the decades that followed.

Goddard's ideas were also met with skepticism and criticism from other scientists, some of whom dismissed them as impractical and impossible. They argued that the technology of the time was not advanced enough to support the kind of research Goddard was proposing, and that the idea of rocket propulsion in space travel was simply too far-fetched to be taken seriously. But Goddard persisted in his research, driven by his passion for understanding the principles of rocket propulsion.

Goddard's early work and education laid the foundation for his groundbreaking contributions to the field of rocket science. His relentless pursuit of knowledge and innovation, coupled with his education in mechanical engineering and physics, would ultimately lead to the launch of the world's first liquid-fueled rocket in 1926.

Chapter 2: Development of the Liquid-Fueled Rocket

Robert Goddard's interest in rocket propulsion began early in his life, and by the time he was in college, he had already begun experimenting with small rockets. However, it wasn't until after he had completed his graduate studies that he began to focus on the study of rocket propulsion in earnest. Robert Goddard conducted a series of experiments firing small rockets and measuring their performance. He had a number of reasons for this.

First, Goddard was interested in understanding the principles of rocket propulsion and improving the efficiency and stability of rockets. By firing small rockets and measuring their performance, he was able to collect data on factors such as thrust, speed and altitude, which helped him understand the mechanics of rocket propulsion and identify areas for improvement. This data would help him understand the missile's behavior and the factors that affect its performance, such as weight, thrust and fuel consumption, which would be critical to the development of more efficient and powerful missiles.

Second, Goddard was interested in exploring the potential of rocket propulsion in space travel, and believed that by conducting experiments on small rockets, he could gather data and insights that would be relevant to the development of larger and more powerful rockets. He believed that by understanding the performance of small rockets he could develop a better understanding of the challenges and opportunities of space rocket propulsion.

Third, Goddard wanted to test his ideas and theories about rocket propulsion and was eager to validate his hypothesis. By conducting a series of small rocket experiments, he was able to test his ideas and theories and collect data to support his claims about the potential of rocket propulsion.

Finally, Goddard wanted to share his findings with the scientific community and gain recognition for his work. By conducting a series of experiments with small rockets and measuring their performance, Goddard was able to collect data that could be used to support his research papers and patents. This helped him gain recognition and credibility within the scientific community and establish himself as a leading rocket science researcher.

One of Robert Goddard key inventions was the development ofa new technique that improved the efficiency and stability of rockets: the use of liquid fuels, especially liquid oxygen and gasoline. Prior to Goddard's work, solid fuels, such as gunpowder, were commonly used in rocketry. However, Goddard recognized the limitations of solid fuels. Solid fuels were inefficient and unstable, creating a lot of smoke and debris. He believed that liquid fuels, such as gasoline and liquid oxygen, would be a much more efficient and stable source of propulsion.

Goddard's use of liquid oxygen and gasoline as rocket fuel was a major breakthrough in rocket propulsion. Liquid oxygen is an oxidizer, which helps to increase the combustion efficiency of the rocket fuel. Gasoline, on the other hand, is a hydrocarbon-based fuel, which can provide a high energy yield, making it an ideal fuel for rockets.

In addition, Goddard's liquid-fueled rocket used a pump-fed propulsion system, allowing greater control over fuel flow and combustion, resulting in a more efficient and stable rocket motor. This system also allowed the missile to be shut down and restarted in flight, which was not possible with solid-fuel rockets.

It is anyone's guess how Robert Goddard came up with the idea of using liquid fuel for rocket propulsion. It is probably a combination of his own experiments, research and theoretical considerations.

Goddard had already conducted experiments with small rockets in his youth. These experiments probably confronted him even then with limitations of solid fuels, such as low efficiency and instability. This may have piqued his interest in finding a more efficient and stable source of propulsion.

In addition, Goddard was of course well versed in physics and mathematics. He had a deep understanding of the principles of thermodynamics and combustion, which would have made him aware of the potential benefits of liquid fuels. Finally, he was of course also aware of developments in his field and ideas of other scientists. Goddard was not alone in thinking about using liquid fuel. Konstantin Tsiolkovsky, a Russian mathematician of Polish descent, already came up with a detailed plan idea in 1895 to build a rocket that ran on liquid fuel. Goddard was not alone in his idea that this should be possible.

Goddard began experimenting with liquid fuels in 1914, and by 1919, he had successfully launched a rocket powered by liquid oxygen and gasoline. This marked a major breakthrough in rocket propulsion, and it paved the way for the development of larger and more powerful rockets that could reach greater altitudes and travel farther distances.

Robert Goddard was not only interested in the propulsion of the rocket but also in how to make the rocket fly in a stable and controlled way. One of the ways Goddard achieved this was through the use of a system he developed called "gyroscopic stabilization." This system used spinning wheels inside the rocket to keep it pointed in the right direction and prevent it from veering off course. Think of it like a bicycle wheel that keeps the bike upright and going straight.

Another way Goddard worked on controlling the rocket's flight was by experimenting with different guidance systems. One example of this is radio-controlled guidance, where Goddard would use radio signals to control the rocket's direction while it was in flight. These innovations were critical in making sure that the liquid fuel rockets, which Goddard developed, could be launched and controlled successfully. So, Goddard's works in stabilization and guidance systems were crucial to make the rocket fly in a controlled way, and to make sure that it reaches its target location.

His relentless pursuit of knowledge and innovation in the field of rocket science would ultimately lead to the launch of the world's first liquid-fueled rocket in 1926, a historic event that would forever change the course of rocket science and space exploration.

The development of liquid-fueled rocket by Robert Goddard was a major achievement in the field of rocket science. His use of liquid fuel, as opposed to solid fuel, marked a major breakthrough in rocket propulsion, as liquid fuels provided a much more efficient and stable source of propulsion. This breakthrough, combined with his advancements in rocket stabilization and guidance systems, would pave the way for the development of larger and more powerful rockets that could reach greater altitudes and travel farther distances.

Chapter 3: Launch of the World's First Liquid-Fueled Rocket

After years of research and experimentation, Robert Goddard was finally ready to launch the world's first liquid-fueled rocket. On March 16, 1926, Goddard and his team gathered at a remote location in Auburn, Massachusetts, to test the new rocket. Goddard chose this location because it provided the privacy and seclusion he needed for his sensitive and dangerous experiments. The isolated location allowed him to conduct his tests without interference or interruption, and also helped minimize the risk of accidents or injury to people or property. In addition, the area was relatively uninhabited, which reduced the likelihood of his experiments being observed or reported. This location in Auburn, Massachusetts became the place where he tested and improved his ideas and designs for liquid fuel rockets, launching the first liquid fuel rocket in 1926.

The rocket, dubbed "Nell," was made of lightweight aluminum and measured just over 9 feet in length. Why the rocket was named "Nell" is no longer known. Maybe the rocket was named after someone on Goddard's team, who knows. However, it is common practice to nickname or codename experimental projects and prototypes, which is how "Nell" was launched. The rocket was powered by a combination of liquid oxygen and gasoline, which would be pumped into the combustion chamber and ignited to create thrust.

No doubt Robert Goddard was quite concerned around the time of the launch. He had to experiment with different types of fuels before finally settling on liquid oxygen and gasoline as the most suitable. The use of highly volatile and hazardous fuels in his experiments posed a significant risk of accident or injury. Another major challenge he should have faced was designing a way to control and stabilize the rocket's flight, as rocket propulsion technology was still in its infancy. And then he needed to come up with an efficient and reliable ignition system for his liquid fuel rocket, a new and untested technology. The weight and size of the rocket were also an issue, as Goddard had to design a rocket that was light and compact enough to be launched, but also strong enough to withstand the forces of flight and propulsion. Finally, Goddard had to find a way to measure and transmit data from the missile during its flight, as the technology for collecting and transmitting remote data was not yet advanced enough.

And those were just the technical problems. Goddard also had concerns in other areas. His research was not exactly widely supported or funded, so he invested a significant amount of his own money in his experiments. A failed test launch could lead to significant financial loss and irreparable reputational damage.

In addition, his work was aimed at developing a rocket capable of reaching space and he may have feared that his experiments would fail to achieve this goal. And Goddard was not alone in working on these ideas. There was serious competition and it was not impossible that someone else would launch a successful rocket before Goddard. In general, early 20th century missile testing was a risky and uncertain business and success was far from assured.

It all worked out well for Robert Goddard. The launch was a historic moment, marking the first successful launch of a liquid-fueled rocket. The rocket's engines ignited and "Nell" flew into the air. It reached a height of 12 meters and covered a distance of 55 meters before landing safely on the ground.

The launch of the world's first liquid-fuelled rocket was a major achievement in rocket science. This breakthrough enabled the development of larger and more powerful rockets that could reach greater heights and travel greater distances. The liquid fuel rockets could generate more thrust than solid fuel rockets and were easier to start and stop, making them more reliable and controllable. In addition, liquid-fuelled rockets have the ability to carry more fuel and oxidizer, meaning they can reach higher altitudes and travel longer distances.

Goddard's breakthrough laid the foundation for future developments in rocket technology, including the use of liquid hydrogen and liquid oxygen as propellants, now commonly used in modern rocket engines. His research into the potential for rocket propulsion in space, which began years earlier, was now validated by the successful launch of the liquid-fueled rocket. It demonstrated the feasibility of liquid fuel rockets and laid the groundwork for the development of larger and more powerful rockets that would eventually lead to space exploration.

Chapter 4: Later Years and Legacy

After the successful launch of the world's first liquid-fueled rocket in 1926, Robert Goddard continued research and experimentation in rocketry. He focused on developing new technologies to improve missile stability and guidance systems, as well as exploring the potential for rocket propulsion in space.

In 1929, Goddard received a patent for a rocket-propelled aircraft, which he believed could be used for high-altitude research and to reach the upper atmosphere. He also began work on a rocket that could reach greater heights and travel greater distances, known as the "Goddard Rocket".

Despite these developments, Goddard faced many challenges in his later years. His work was often met with skepticism and criticism from the scientific community, and he struggled to secure funding for his research.

In addition, during the Great Depression, funding for scientific research was limited anyway, making it difficult for Goddard to continue his work.

Even after Robert Goddard successfully launched his rocket "Nell" in 1926, criticism of his research persisted. The lack of understanding of his ideas, skepticism from the scientific community and the media, limited public awareness of his work, and a lack of government funding continued to plague him. Goddard's ideas about rocket propulsion and space travel were ahead of their time, and many people in the scientific community and the media did not fully understand or appreciate the potential of his work. What didn't help was the remote location in Auburn. The experiments received little publicity because of this, which meant that the general public and media had limited access to information and visual evidence of the success of Goddard's experiments.

In 1935 Goddard moved to Roswell, New Mexico. In 1935 Goddard moved to Roswell, New Mexico. This was a strategic decision that enabled him to continue his pioneering research into rocket technology with better facilities and conditions. One of the main reasons for the move was the weather, as Roswell's clear skies, low humidity, and mild temperatures provided an ideal environment for year-round missile testing. In addition, like Auburn, Roswell's remote location provided the privacy and seclusion Goddard needed for his sensitive and dangerous experiments, allowing him to conduct his tests without interference or interruption and minimizing the risk of accident or injury to people or property. could minimize.

Another important factor in Goddard's move to Roswell was the funding he received from the US government. Roswell was chosen as the location for the new research facility due to its remote location and favorable weather conditions. Its location in the desert southwest, far from major population centers, also made it an ideal location to launch rockets into space without posing a threat to people or property in the event of a launch failure.

In the last ten years of Robert Goddard's life, he continued to work on rocket technology and space exploration. However, his health began to decline and he suffered from serious heart disease and diabetes. Despite his health problems, Goddard continued to develop new rocket designs and propulsion systems, and he received funding from the United States government for his research.

In the early 1930s, Goddard received funding from the Smithsonian Institution to support his research and experiments. He also received funding from the Guggenheim Foundation to support his work on high-altitude rocket research. Goddard was able to launch several rockets during this period, including one that reached an altitude of 2200 meters in 1935.

In addition to his rocket research, Goddard also worked on the development of other technologies. He invented a device to stabilize aircraft, which was patented in 1931, and he also worked on the development of a remote control system for aircraft.

Goddard's health continued to decline and he died on August 10, 1945, at the age of 62. Despite the criticism and skepticism he faced during his lifetime, Goddard's pioneering work in rocket technology laid the foundation for space exploration.

Robert Goddard's legacy lives on as his pioneering work in rocketry continues to inspire scientists and engineers around the world. His contributions to rocket science, including his development of the liquid-fuelled rocket, his work on rocket stabilization and guidance systems, and his research into the potential for rocket propulsion in space, have played a vital role in the development of modern space technology.

In recognition of his contributions to the rocketry field, Goddard has been honored with numerous honors and awards. The Goddard Space Flight Center, operated by NASA, is named after him, as is the Goddard Planetarium in Worcester, Massachusetts, where he grew up. In addition, the American Institute of Aeronautics and Astronautics has created the Robert H. Goddard Memorial Trophy, which is awarded annually to an individual or organization that has made significant contributions to the rocketry and aerospace field.

In conclusion, Robert Goddard's later years and his legacy remain an inspiration to all future scientists and engineers. His relentless pursuit of knowledge and innovation in rocket science and space exploration has played a vital role in the development of modern space technology, and his contributions to rocket science will be remembered and celebrated for years to come.

Chapter 5: The legacy of Robert Goddard in the rest of the world

Robert Goddard is widely regarded as a pioneer of rocket technology and space exploration, both in the United States and abroad. His ideas and inventions were ahead of their time, and Goddard's work is recognized as a seminal influence on the development of rocket technology and space exploration. Today, Goddard's contributions to rocket technology are recognized by many countries and his work is studied in schools and universities around the world.

Goddard is especially celebrated in countries with strong space programs, such as the US, Russia, China, Japan, India and Europe, and his work is often studied as part of the history of rocketry and space exploration in these countries. Many of Goddard's patents and scientific papers have been translated into foreign languages and are widely read and studied abroad.

Let's see what is being watched in the countries mentioned to Robert Goddard

Russia

During the Soviet era, Goddard's work was studied and acknowledged by Russian scientists and engineers, and it was an inspiration for the development of the Soviet rocket program. Many of Goddard's patents and scientific papers were translated into Russian and were widely read and studied in Russia.

After the collapse of the Soviet Union, Goddard's work continues to be acknowledged and studied in Russia. His contributions to rocket technology are considered to be important in the history of space exploration and science. Goddard's work is studied in schools and universities and is an important part of the curriculum in the field of aerospace engineering and space technology.

China

During the Cold War era, China's access to foreign technology was limited, and information about Goddard's work was not widely available. However, after the establishment of diplomatic relations with the United States in 1979 and the opening of China to the outside world, Goddard's work began to be more widely known and studied in China.

In recent years, China's space program has developed rapidly, and Goddard's work is studied and acknowledged by Chinese scientists and engineers. His contributions to rocket technology are considered to be important in the history of space exploration and science. Goddard's work is studied in schools and universities and is an important part of the curriculum in the field of aerospace engineering and space technology.

Japan

Japan's space program has been developing since the 1950s, and Goddard's work has been studied and acknowledged by Japanese scientists and engineers. His contributions to rocket technology are considered to be important in the history of space exploration and science. Goddard's work is studied in schools and universities and is an important part of the curriculum in the field of aerospace engineering and space technology.

In recent years, Japan's space program has grown significantly, and Japan has developed its own rockets and launched several satellites into space. Goddard's work continues to be an inspiration for Japanese scientists and engineers in the field of rocket technology and space exploration.

India

India's space program began in the 1960s, and Goddard's work has been studied and acknowledged by Indian scientists and engineers. His contributions to rocket technology are considered to be important in the history of space exploration and science. Goddard's work is studied in schools and universities and is an important part of the curriculum in the field of aerospace engineering and space technology.

In recent years, India's space program has grown significantly, and India has developed its own rockets and launched several satellites into space. Goddard's work continues to be an inspiration for Indian scientists and engineers in the field of rocket technology and space exploration. India's space research organization, Indian Space Research Organization (ISRO) has been able to achieve many milestones in space exploration, and Goddard's work continues to be an inspiration for them.

Europe

European countries have been involved in space research and exploration since the early days of the space age, and Goddard's work has been studied and acknowledged by European scientists and engineers. His contributions to rocket technology are considered to be important in the history of space exploration and science. Goddard's work is studied in schools and universities and is an important part of the curriculum in the field of aerospace engineering and space technology in many European countries.

In recent years, European countries have developed their own space programs, and Goddard's work continues to be an inspiration for European scientists and engineers in the field of rocket technology and space exploration. The European Space Agency (ESA) has been able to achieve many milestones in space exploration and Goddard's work continues to be an inspiration for them.

Chapter 6: Robert Godard in the education program

The work of Robert Goddard is widely studied and acknowledged as an important part of the history of rocket technology and space exploration. As such, his work is included in the curriculum of many programs that teach aerospace engineering and space technology.

Some examples of programs that teach from the work of Robert Goddard as standard include:

- Aerospace engineering programs at universities: Goddard's work is studied as part of the history of rocket technology and space exploration in many aerospace engineering programs at universities around the world. Students learn about Goddard's contributions to the field and how his ideas and innovations laid the foundation for modern rocket technology.

- Space science programs at universities: Goddard's work is also studied in many space science programs at universities around the world, where students learn about the history of space exploration and the key figures who have contributed to the field.

- Professional training programs for NASA and other space agencies: Goddard's work is also studied as part of the training programs for engineers and scientists working for NASA and other space agencies around the world. These programs cover the history and current state of the field of rocket technology and space exploration, and Goddard's work is an important part of the curriculum.

- High school education: Goddard's work is also included as a part of the curriculum in some high school education programs, where students learn about the history and science of space exploration and rocket technology.

Overall, Goddard's work is widely studied and acknowledged as an important part of the history of rocket technology and space exploration.

Chapter 7: Robert Godard and NASA

A major player in space technology is NASA, the National Aeronautics and Space Administration. NASA also recognizes Robert Goddard as one of the pioneers of modern rocket technology and space exploration. Goddard's work in the early 20th century laid the groundwork for many of the technologies and concepts used by NASA today, and his contributions to the field are widely recognized and celebrated by the agency.

NASA has named several facilities and projects after Goddard, including the Goddard Space Flight Center in Greenbelt, Maryland, one of the agency's main research centers, and the Goddard Institute for Space Studies in New York City, which conducts research in climate and Earth science . Goddard's work is also studied by NASA scientists and engineers as part of their training and education.

NASA also maintains a website dedicated to Goddard's work, where you can find a wealth of information about his life, contributions and legacy. The website contains Goddard's biographical information, a history of his work, and a list of his patents, publications, and photographs.

Chapter 8: The Private Life of Robert Goddard: A Look Beyond the Rocket Scientist

Of course, when we look at Robert Goddard, we first see the brilliant scientist who made a breakthrough in rocket technology. But of course Robert is also a human being. In this chapter we look at what we can learn about Robert Goddard as a person.

As we wrote at the beginning of this book, Robert Goddard was born on October 5, 1882 in Worcester, Massachusetts.

Goddard attended Worcester Polytechnic Institute, where he received a degree in physics in 1908. He then attended Clark University, earning a master's degree in 1909 and a doctorate in physics in 1911. During his time at Clark, Goddard met Esther C. Kisk, whom he would later marry in 1924. Esther was a supportive partner throughout Goddard's career, helping him with his research and experiments and managing his business affairs. She also served as his secretary, typist and editor. They kept their marriage private and not much is known about their personal lives together, but it is known that Esther was a loyal companion to Goddard and helped him through his personal and professional difficulties. Ester died in 1966. As far as is known, no children were born from this marriage.

Not much is known about Robert Goddard's personal friendships as he was a relatively private person. He was known for being close to his family, especially his younger brother Richard. Goddard's work and private life were closely intertwined and he devoted much of his time to his research and experimentation. He did maintain some professional relationships with other scientists and engineers, but it is not clear to what extent these were close personal friendships. Goddard was in correspondence with some of the notable figures of his time, such as Charles Lindbergh, who took an interest in Goddard's rockets and helped him raise money for them. Goddard also corresponded with Harry Guggenheim, a philanthropist who provided significant funding for Goddard's research.

Goddard's personal life has been marked by a number of challenges. He suffered from ill health throughout his life and was often bedridden with illnesses such as tuberculosis and pneumonia. He also struggled with depression and feelings of isolation, which were likely exacerbated by his lack of acceptance and support from the scientific community.

Despite these challenges, Goddard remained committed to his work and continued to make pioneering contributions in the field of rocketry. He passed away on August 10, 1945 at the age of 62, leaving behind a legacy that continues to inspire future generations of scientists and engineers.

Robert Goddard has made many statements about his work and beliefs throughout his career, but some of his most notable statements you can read on the following pages.

"It is difficult to say
what is impossible,
because yesterday's
dream is today's hope
and tomorrow's
reality."

"I am convinced that space travel will one day become as common as crossing the ocean."

"It is the nature of every human being to face with composure any danger that is not really present."

"Every vision is a joke until the first man realizes it; once realized, it becomes commonplace."

These statements reflect Goddard's confidence in his work and the capabilities of rocket technology and space exploration. They also show how firmly and unreservedly he believed in what he was doing, despite the skepticism and lack of funding he faced in his career.

Conclusion

Robert Goddard, known as the "Father of Modern Rocketry," was a pioneering American engineer, professor, physicist, and inventor who is credited with creating and building the world's first liquid-fueled rocket. Throughout his life, Goddard was driven by his passion for understanding the principles of rocket propulsion and his relentless pursuit of knowledge and innovation. His early work and education laid the foundation for his groundbreaking contributions to the field of rocket science, which culminated in the launch of the world's first liquid-fueled rocket in 1926, a historic event that would forever change the course of rocket science and space exploration.

Goddard's work on liquid-fueled rockets was a significant breakthrough in rocket propulsion, as liquid fuels provided a much more efficient and stable source of propulsion. This breakthrough would pave the way for the development of larger and more powerful rockets that could reach greater altitudes and travel farther distances. Additionally, Goddard's research on the potential for rocket propulsion in space, validated by the successful launch of the liquid-fueled rocket, marked the first step towards the exploration of space.

Despite facing skepticism and criticism from many in the scientific community, Goddard persisted in his research and experimentation, driven by his passion for understanding the principles of rocket propulsion. Despite the challenges he faced, Goddard's legacy lives on today, as his pioneering work in rocketry continues to inspire scientists and engineers around the world.

In recognition of his contributions to the field of rocketry, Goddard has been honored with numerous awards and accolades. The

Goddard Space Flight Center, operated by NASA, is named in his honor, as is the Goddard Planetarium in Worcester, Massachusetts, where he grew up. Additionally, the American Institute of Aeronautics and Astronautics has established the Robert H. Goddard Memorial Trophy, which is awarded annually to an individual or organization that has made significant contributions to the field of rocketry and astronautics.

In conclusion, Robert Goddard's pioneering contributions to the field of rocket science and space exploration have played a crucial role in the development of modern space technology. His relentless pursuit of knowledge and innovation will continue to be remembered and celebrated for years to come, and he will always be remembered as the "Father of Modern Rocketry".

Appendix

"Rocket Man: Robert Goddard and the Birth of the Space Age" by David A. Clary

"Robert Goddard: Pioneer of Space Flight" by Milton Lehman

"Robert Goddard: Father of the Space Age" by Milton O. Thompson

"The Papers of Robert Goddard" edited by Milton W. Rosen

"Goddard: Pioneer of Space Research" by Lloyd S. Swenson Jr., James M. Grimwood, and Charles C. Alexander